Scented Herb Papers

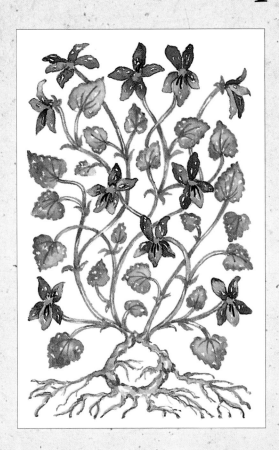

By viewing nature, nature's handmaid art,
Makes mighty things from small beginnings grow.

Dryden

For my beloved parents.

Scented
Herb Papers

*How to use natural scents and colours in
hand-made recycled and plant papers*

Polly Pinder

SEARCH PRESS

First published in Great Britain 1995

Search Press Limited
Wellwood, North Farm Road,
Tunbridge Wells, Kent TN2 3DR

ISBN 0 85532 789 8

If you have difficulty in obtaining any of the materials or
equipment mentioned in this book, please write for further
information to the Publishers.
Search Press Limited
Wellwood, North Farm Road
Tunbridge Wells, Kent TN2 3DR
England

The author would like to thank Messrs G. Baldwin of London
for supplying many of the essential oils used in this book

Printed in Spain by Elkar S. Coop, Bilbao 48012

Contents

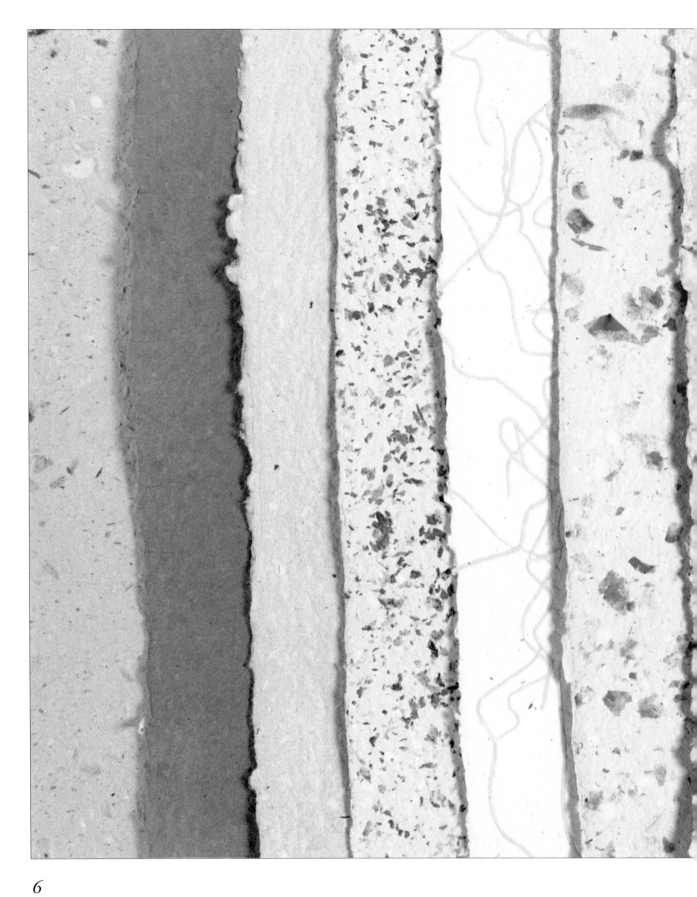

Introduction

The fairest things have fleetest end,
Their scent survives their close.

These words by the Victorian poet
Francis Thompson sum up the idea
behind this book: to capture the
evocative perfume of the summer
garden in hand-made paper.

Making paper, like all creative
ventures, releases the imagination
and offers a great deal of satisfaction.
In particular, there is plenty of
pleasure in store for the senses –
swirl your hands through a bowl full
of soft paper pulp, feel the crisp
firmness of a piece of newly made
paper, smell the lingering fragrance
of a sheet of lavender paper, and
enjoy the romantic effect of scattered
petals embedded in a piece of rose
paper.

The idea of creating something
beautiful with your own hands to
give to someone you love, or to keep
for yourself, is a truly lovely one.

Materials and equipment

This is not a craft that requires a vast amount of equipment. Most of the items you will need to make the paper will be in your kitchen, while many of the plants needed to perfume and tint the paper may already be growing in your garden.

Our postage-stamp of a garden has only a small lawn, but every available inch of wall is clad in honeysuckle and jasmine, and just outside the kitchen door we grow many varieties of sweet-smelling herbs in clay pots .

Most of the pulp made for the papers in this book was produced from shredded photocopy and computer paper. Many large offices and administrative departments accumulate vast quantities of shredded paper which are normally thrown away. You could arrange to recycle this for your paper-making. Otherwise, use old newspapers and any other uncoated papers, torn into narrow strips.

Shredded paper may come in different colours, which will obviously affect the colour of your final paper – in most cases this is an advantage as it introduces variation without having to use dyes (see the examples opposite, and those on pages 26, 42 and 52). The shredded paper will also contain printing ink, which sometimes produces a grey tint in the final paper. You can reduce this by squeezing the pulp in a piece of nylon netting after it has been through the food processor, then rinsing it thoroughly in a colander lined with netting.

Left: Sheets of paper made from shredded photocopy paper.

Opposite: Papers made from coloured shredded office paper. The delicate peach (bottom right) is made from a combination of all three.

The following list of materials and equipment will make about sixty-five sheets of paper, each approximately 125 x 175mm (5 x 7in). This may seem a large amount to begin with, but it will give you a good opportunity to feel your way through the process of paper-making, to make mistakes and then to start experimenting and putting your own ideas into practice. Once the pulp has been made it will keep in an uncovered bucket almost indefinitely (as long as no fresh vegetable juice or matter has been added). You can then have a paper-making session whenever the mood takes you, without having to get the food processor out again.

- Two 10-litre (2-gallon) plastic buckets
- Enough shredded paper to fill one of the buckets three-quarters full when pressed down firmly
- Soup ladle (not essential)
- Large plastic colander and piece of nylon netting approximately 50cm (20in) square for washing the pulp
- Food processor
- Washing-up bowl approximately 30 x 40cm (12 x 15in)
- Smaller bowl or basin
- Shallow plastic tray approximately 30 x 40cm (12 x 15in)
- Two packets of all-purpose kitchen cloths
- Lots of newspapers
- Deckle and mould approximately 15 x 20cm (6 x 8in) – available at most art and craft shops
- Two pieces of laminated chipboard approximately 30 x 40cm (12 x 15in)
- Four G-clamps – available at DIY stores
- Cloths for mopping up
- Small jug

Plastic colander and piece of nylon netting

All-purpose kitchen cloths

Bowl

Jug

Deckle and mould

Plastic bucket

Shredded paper

Washing-up
bowl

Soup ladle

Cloths

Four G-clamps

Shallow plastic tray

Laminated chipboard

BRAUN

Food processor

Newspapers

11

Making recycled paper

The pulp

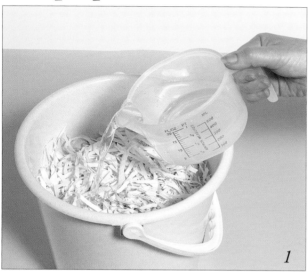

1. Add warm water to the bucket of shredded paper. Keep pressing it down and adding water until the bucket is a third full – the paper will sink down almost immediately.

2. Leave the paper to soak, stirring occasionally with your hand or a wooden spoon. The paper will break up more easily if it is left overnight.

3. Ladle three scoopfuls of pulp into the food processor and process it for about thirty seconds. Stop, then repeat the processing until the strips of paper have completely disintegrated. Empty the processed pulp into a bowl, and continue processing until you have used all the raw pulp.

4

5

4. Wash the processed pulp to reduce the grey tint caused by the printing ink. After washing, the pulp can be processed again in order to create a smoother paper.

5. Place the washed pulp in the second bucket and just cover it with water. The pulp can be used immediately or left, as long as it is not covered with a lid, until you are ready for a paper-making session.

Couching

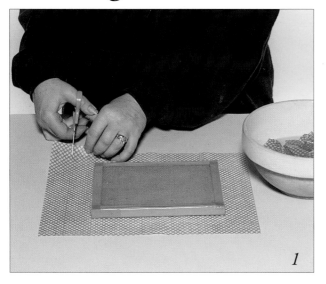

1. Cut the all-purpose cloths roughly 5cm (2in) larger all round than the mould. Put twelve pieces aside to use as padding, together with two newspapers – to be put between the boards to absorb excess water when pressing the paper. Put the remaining pieces in a small bowl with some water.

3. Add enough water to wet the newspapers thoroughly. Lift each layer to ensure that the water has penetrated right through. Smooth out any cockling to make a shallow-domed mound and then tip away the excess water.

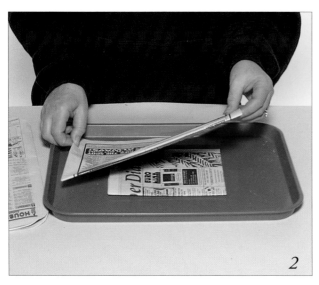

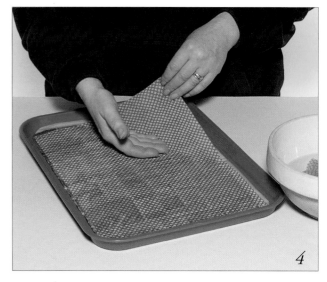

2. Make the couching pad (on which your sheets will be laid when the deckle is removed) from three full sheets of newspaper. Fold the first three times, the second two times and the third once. Lay these in the centre of the tray; first the small one, then the medium one, and then the large one.

4. Squeeze out one of the cloths in the basin and lay it carefully on top of the newspapers, smoothing out any creases.

Moulding

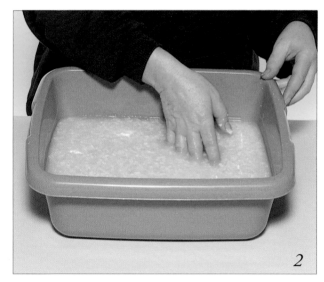

1. Scoop three handfuls of the processed pulp into the bowl. This will make between twelve and fifteen sheets of paper.

2. Add enough water to fill the bowl three-quarters full. Gently stir the water so that the pulp is evenly distributed.

3. With the net uppermost, place the deckle on top of the mould.

4. Hold the deckle and mould firmly together and slide them, at a slight angle, into the bowl.

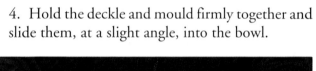

5. Lift the deckle and mould up, keeping them as horizontal as possible, and allow excess water to drain through the mould back into the bowl.

6. Carefully lift the deckle off the mould and put it to one side.

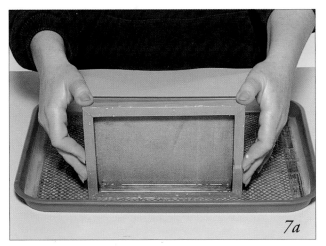

7a

8

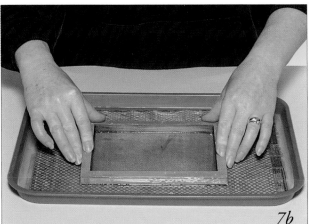

7b

8. Take another cloth from the basin, squeeze it out and, starting from one corner, carefully lay it on top of the sheet of paper, making sure there are no creases.

7c

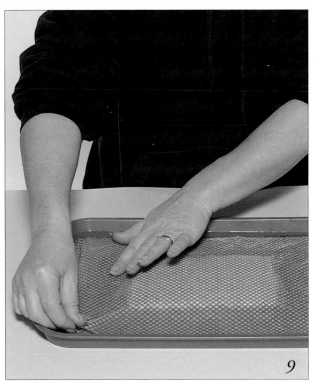

9

7. Carefully stand the mould vertically on the edge of the couching mound with the paper facing you. Now lay it down in a rolling motion, pressing it gently on to the couching mound and lifting it up at the opposite side. The sheet of paper will transfer to the couching mound.

9. Continue to make sheets of paper, laying them down on the couching mound, until there is very little pulp left in the water. Pour any accumulated water from the tray back into the bowl. Cover the last sheet with a squeezed-out cloth as before.

Pressing and drying

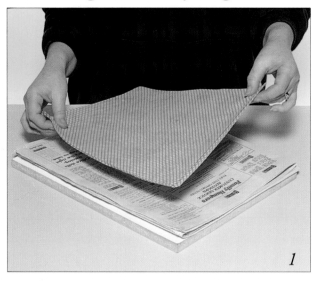

1

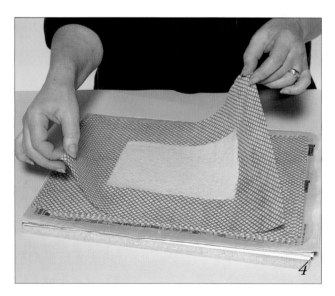

3

1. Prepare the pressing boards. Lay a folded newspaper on one of the boards and then lay six of the cloths centrally on top.

3. Attach the G-clamps and slowly begin to tighten them. Water will begin to run out from between the boards so hold them over the bowl – or be prepared to mop up afterwards! When the clamps have been fully tightened, leave for a few minutes and then dismantle the press.

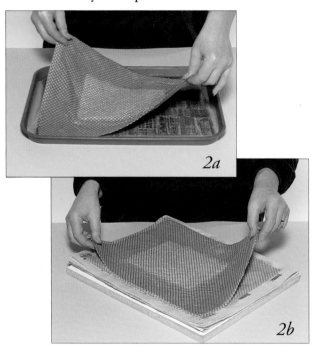

2a

2b

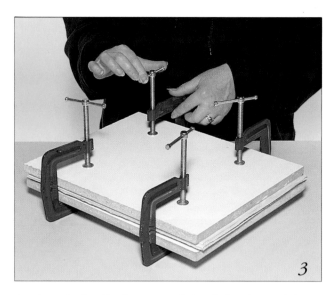

4

2. Take two opposite corners of the bottom couching cloth and then carefully lift everything off the couching mound on to the prepared board. Smooth over the surface and cover with the remaining cloths, another newspaper and the other board.

4. Remove the top layer of newspaper and padding and the top couching cloth to expose the top sheet of paper. Carefully peel away each couching cloth with its accompanying sheet of paper.

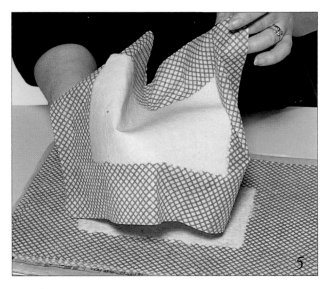

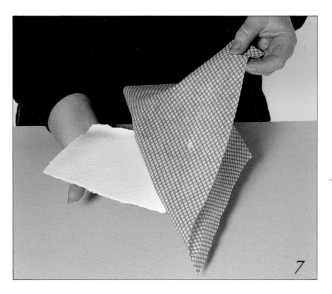

5. Do not worry about bending the sheet of paper. By now the fibres will have joined together to form a flexible sheet. However, the sheets are still wet, so do not be tempted to remove the sheet from the couching cloth.

7. When the sheets of paper have dried (it usually takes about sixteen hours), peel away the couching cloths.

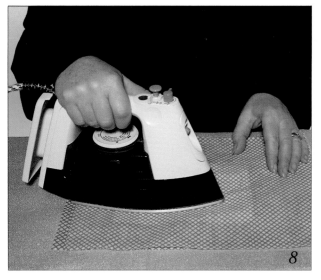

6. Dry the paper on the couching cloth. The usual drying method is to lay the cloth on newspaper. This is fine but as the newspaper dries it also wrinkles, and so does the handmade paper. I have found the best way to obtain beautifully flat sheets of paper is to leave them to dry on the carpet.

8. Iron each sheet of the paper with a steam iron between two dry cloths, then leave the sheets of paper overnight under a pile of books or magazines. If the carpet method is inconvenient and you have used newspaper, iron the sheets of paper while they are still slightly damp.

Now you have learnt how to make recycled paper, you can start the really creative part! A walk in the country, a tour of the garden or a trip to the shops will throw up dozens of ideas for colouring, scenting and embellishing your paper – from roses and geraniums to sage and rosemary. You can even make paper from dead leaves
(see page 56).

Sharon fruit

Chilli

Thyme

Mango

Sweet cicely

Lime

Red pepper

Lychee

Sage

Rosemary

Origanum

Rose

Viola

African violet

Primrose

Autumn leaves

Bamboo

Straw

Hay

Eucalyptus

Geranium

Ivy

Bracken

21

Adding natural colour

You can add many bright colours in the form of inks and dyes to the pulp to create an interesting array of papers, but there are also numerous natural ways to achieve subtle and understated colours. Tea leaves, onion skins, blackberries and beetroot all provide good colours.

Obviously, the strength of any colour will depend on the amount of liquid added to the pulp in the bowl, and that will depend on the amount of raw ingredients used. If you add a cupful of liquid to the pulp, the colour will be subtle. However, the colour will be stronger if the boiled juice of vegetable matter (instead of clean water) is used to bring the pulp to the right consistency at stage 2 on page 15.

Tea leaves

I was able to be excessively generous with tea leaves when making dye for the paper on pages 43, 45 and 46. Several years ago my daughter bought me a large caterer's caddy full of tea which for some reason I had never used. I boiled most of it in a jam pan full of water and the result was a very strong dye and a lovely tea scent which floated through the house.

Onion skins

The outer skins of onions will give you shades ranging from delicate cream to deep orange. Make sure the skins are perfectly dry before storing them in an airtight jar. To get the cream shade, add the boiled juice from a pan full of onion skins to the pulp. You can make deeper shades by dipping the sheet of paper into a shallow dish of concentrated onion-skin juice, then hanging the paper on a makeshift line with paperclips to dry.

Blackberries

Blackberry sediment is made by scraping fresh blackberries through a sieve. When you add this to

Tea leaves

Onion skins

Blackberry sediment

Natural dyes can be found in a variety of plants and fruits

Beetroot juice gives a pale-pink tint to the paper, with a deeper pink on the deckled edges.

the pulp bowl you will find it gives a speckled blue/grey tint to the paper. Blackberry juice is made by boiling the fruit in as little water as possible, then straining the juice and berries through nylon netting or muslin. This produces a very pale pink.

Beetroot
When beetroot juice is added to the pulp bowl, the white pulp becomes a vibrant pink. Unless you use a mordant (fixing agent), most of the colour will be lost during the drying process, but a subtle hint of pink remains, with a slightly stronger shade lingering in the deckled edges.

Storage of pulp
The juice or pulp from any vegetable matter will go off or start to ferment after a day or two. It will keep a little longer in the fridge and for months in the freezer. If you have to stop during the middle of a paper-making session and cannot get back to it for some time, the best way to preserve, for example, pulp containing lavender flowers is to strain it through nylon netting, squeeze it until all the juice is extracted, then form the damp pulp into a ball and leave it to dry naturally, wrapped in an all-purpose cloth. Freeze the juice, as this will contain some of the perfume. You can reconstitute the pulp at a later date.

I dare say that I have not explored all colour ideas, but the kitchen garden (or in my case the greengrocer's shop) and our hedgerows offer endless possibilities.

Scenting paper

The easiest and most natural way is to add a handful of sweet-smelling herbs, preferably dried, to the pulp bowl. This will not only perfume the paper but also add texture and sometimes colour.

Adding a handful of dried herbs to the pulp bowl produces a scented and textured paper.

Although many herbs and spices have a particularly strong smell when they are fresh or dried, some seem to lose it during the paper-making process. I was astonished to find that a very old – about fifteen years old, in fact – packet of rose pot pourri added to the pulp bowl still retained its delicate perfume after having been incorporated into the paper.

However, after much experimentation I have discovered that the most practical way of enhancing a weak scent is to spray the paper with a pure essential oil (see page 30).

Left: **Mint paper**
This was made from a mixture of pink, orange and white shredded office paper. I then added a handful of dried mint to the pulp.

Opposite: **Lavender paper**
In addition to dried lavender I also added a cupful of blackberry sediment to the pulp to give a trace of colour to the paper (see also page 22).

Citrus paper
Thin slivers of orange and lemon peel were added to the pulp, some of which floated to the top.

O love what hours were thine and mine,
In lands of palm and southern pine;
  In lands of palm, of orange blossom,
Of olive, aloe, and maize and vine.

                                    Tennyson

Spraying essential oil

Bottles of essential oil can be bought at many chemists and herbalists. You can spray it on to the paper using a simple diffuser (available from most art and craft shops). The spraying can be done while the paper is still slightly damp, on sheets of newspaper, or, again on newspaper, after the paper has been ironed.

Some oils, patchouli for example, are rather dark. They may affect the colour of your paper but because the paper is absorbent the colour spreads evenly, and is actually enhancing rather than detrimental.

It would be nice to be able make your own essential oil – unfortunately, both the equipment and the space needed for such an undertaking is prohibitive for all save those who want to embark on a commercial venture.

You can spray pure essential oil on to the paper to enhance and prolong the scent.

A selection of essential oils

Amber	Ginger	Orange	Spearmint
Angelica	Grapefruit	Origanum	Spike lavender
Basil	Hyssop	Palmarosa	Thyme
Bay	Jasmine	Parsley	Verbena
Bergamot	Juniper	Patchouli	Vetivert
Cardamom	Lavender	Peppermint	Ylang ylang
Caraway	Lemon	Pine	
Cedarwood	Lemongrass	Rose geranium	
Chamomile	Lime	Rose maroc	
Cinnamon	Mandarin	Rosemary	
Citronella	Marjoram	Rosewood	
Clary sage	Melissa	Sage	
Clove leaf	Myrrh	Sandalwood	
Coriander	Neroli		
Cypress	Nutmeg		
Fennel			
Geranium			

Sage paper (top)
To create the sage paper, I added dried and freshly chopped sage leaves to the pulp bowl, with a little cabbage sediment from boiled savoy cabbage leaves to give the subtle green tint.

Cinnamon paper (bottom).
For this I added a mugful of onion-skin juice and broken cinnamon sticks to the pulp in the bowl before making the sheets. Initially the scent was strong but it faded as the paper dried, so I sprayed some essential oil of cinnamon on to it after ironing.

Rose paper
I added a fifteen-year-old packet of rose pot pourri to the bowl of pulp to produce this beautiful paper — which still retains the exquisite scent of roses.

Hand-made boxes containing fragrant roses

I made the round box from a cardboard tube (the kind used for storing maps or posters) then covered and lined it with paper which had been dipped in a tray of beetroot juice. I trimmed the edges of the box with florist's tape. The rose petal on the lid was inlaid just before couching and the other was carefully attached, using PVA glue, when the box was completed.

The square box, made from maroon mounting board, was covered with rose paper. The two roses were cut from the same pattern but the pink one has torn edges which I dabbed with extra beetroot juice. I sprayed the pink rose with essential rose oil – the other retained the scent from the pot pourri which I had added to the pulp.

Left: **A round box decorated with rose petals.**

Opposite: **Two hand-made-paper roses and a square box covered with rose paper.**

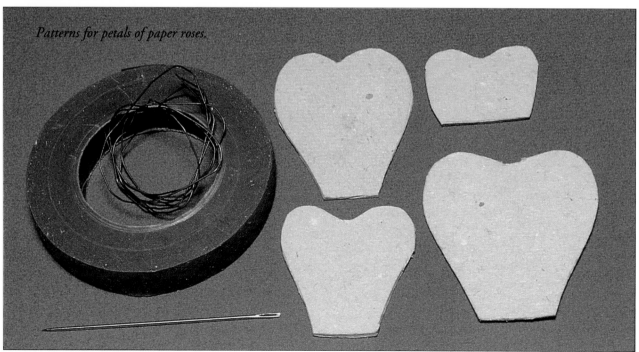

Patterns for petals of paper roses.

Inlaying natural materials into paper

Almost any small piece of foliage can be inlaid and become part of the paper. This occurs naturally when certain dried herbs are added to the pulp bowl; most of the pieces become integrated with the pulp while the remainder float to the top and consequently lie on, rather than in, the paper.

To inlay, either scatter or carefully position leaves or petals on the pulp after you remove the deckle and just before you turn the mould on to the couching cloth. Dry the sheet of paper in the usual way but do not iron the side in which the foliage is inlaid. If the foliage starts to come away from the paper, smear it very gently with a tiny dab of PVA glue then carefully press it back. There will be an indent in the paper into which it will fit snugly again.

If the inlaid plant starts to come away from the paper, stick it back on with a tiny dab of PVA.

Inlaying sage leaves just before the pulped sheet is turned on to the couching cloth.

Sprays of garden conifer on a paper dyed with strong tea.

Dark behind it rose the forest,
Rose the black and gloomy pine trees,
Rose the firs with cones upon them.
Longfellow

The bud may have a bitter taste But sweet will be the flower.
Cowper

Sweet-cicely leaves, inlaid on a sheet of tea-dyed paper (made with a weak solution of tea).

Groves whose rich trees wept
odorous gums and balm,
Others whose fruit burnished
with golden rind
Hung amiable.

Milton

There's fennel for you,
and columbines; there's
rue for you; and here's
some for me.

Shakespeare

Orange peel (top) and the flower
head of bronze fennel (bottom),
inlaid just before couching.

38

And I will make thee beds of roses
And a thousand fragrant posies.

Marlowe

A single inlaid rose petal on paper coloured with beetroot.

Green grow the rushes O.
Burns

Club rush (Scirpus cernuus, *a house plant) embedded in recycled computer paper.*

Above: **Gift tags and cards**
Here I have used salad burnet leaves and large daisy petals (top); and carnation petals (bottom).

Opposite: **Floral writing paper**
These scented papers are decorated with marigold petals (top), primula petals and a fuchsia (centre); and hyssop flowers (bottom).

Exotic stationery
*The paper was made from various combinations of orange, pink and white pulp, inlaid
with pressed marigold petals then sprayed with essential oils of mandarin and patchouli. The
envelopes were cut from the pattern of an existing envelope, reduced in width to fit the box.*

Geranium drawer liners Dyed with tea, these liners were inlaid with whole pressed rose-geranium leaves and torn fresh ones. I added a spray of rose-geranium oil after ironing.

Scented gift wrap and tags
Both of these sheets of gift wrap were inlaid with paper, the white with fine strips of hand-made paper and the fawn (tea-dyed) with tiny squares of yellow duplicating paper liberally scattered over the surface. The papers were scented with lemon and rose oils.

Scented greetings cards

The paper flower pot (dyed with onion skins) and pressed parsley were inlaid into white paper, which was then mounted on to paper dyed with tea. I sprayed the card with essence of parsley and attached a packet of parsley seeds to the inside.

The tea card, with a small packet of tea inside, was sprayed with tea-tree essence. I made the paper for the card by adding tea and tea leaves to concentrated pulp. The empty mug of tea was cut from white hand-made paper which had a line of tea leaves sprinkled across it just before couching, then I stuck both it and the little mat on the card with PVA glue.

Perfumed book-marks (opposite)

I inlaid pressed feverfew petals, sprigs of hyssop, fuchsia leaves, and a strip of hand-made paper to make these book-marks. I then sprayed them with oils of sandalwood, hyssop and ylang ylang.

46

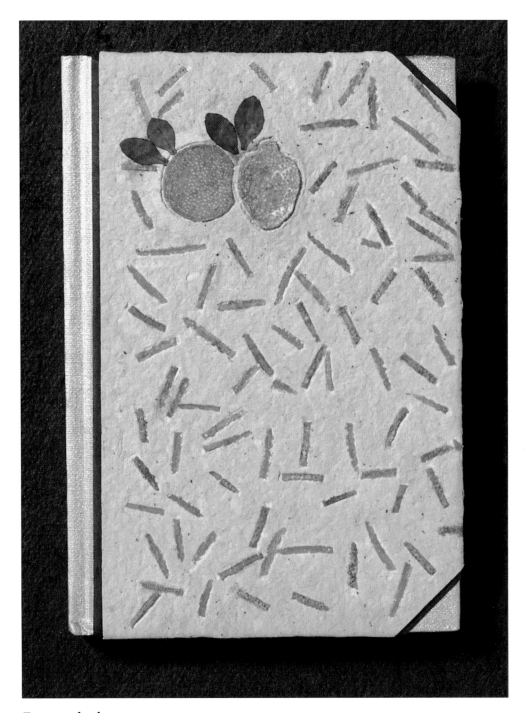

Fragrant book covers

These are really an adornment rather than a protection. The citrus cover, inlaid with tiny slivers of orange and lemon peel, pennyroyal mint leaves and fruit shapes cut from the rind, was sprayed with orange and lemon essential oils. Before covering the book I painted it back and front with enamel gold paint and positioned inset strips of dark-green paper. The fuchsias were inlaid, and as they have no perfume themselves I sprayed the paper with jasmine oil. The covers were attached with PVA glue.

Fragrant lampshade
I made this simple shade from sheets of hand-made paper and pressed ferns, then sprayed it with cedarwood and musk oils. The fragrance is accentuated by the warmth of the light bulb. I attached the ferns to the paper using PVA glue, but they could easily have been inlaid just before couching. Each section of the shade was glued to the frame with PVA.

Scented candle-shades

Scented candle-shades are a pretty addition to any table-setting and are very simple to make. Cut the bottom deckle edge from a sheet of hand-made scented paper, then stick a strip of double-sided adhesive tape down one side and join the two edges together. You should place the candle inside a glass jar which is taller than the flame and, for safety, it is wise not to leave lighted candles unattended.

Sage candle-shade

I added dried and torn fresh sage leaves to the pulp, together with the juice from boiled savoy cabbage leaves to give a hint of green. The natural scent was strong enough not to require additional essence.

Cinnamon candle-shade

I added a handful of broken cinnamon bark to the pulp, along with a mug of onion dye to give the subtle yellow tint. As the broken spice was not fragrant enough in itself I sprayed the paper with cinnamon essential oil before making it into a shade.

Minty place setting
This paper was made from orange and pink office paper with a handful of dried mint added to the pulp. The scent was fairly strong but I enhanced it by spraying on oils of peppermint and spearmint. The napkin ring is a strip of the paper wrapped round a piece of cardboard tube.

Impressing objects into paper

You can impress foliage with enough depth and textural interest into the paper just before you remove the deckle. The foliage can be taken out before or after ironing. You can also use other suitable objects to achieve decorative effects, such as plastic doilies, string, bent wire, or pieces of cork cut into different shapes.

Detailed impressions can be made in the paper using suitable foliage and other objects. Here I have used a piece of conifer.

Using cork
I have cut decorative shapes from a cork coaster and used these to impress motifs in the paper.

Using string
I coiled pieces of string, stuck them with PVA glue on to round pieces of card, then pressed them into wet paper.

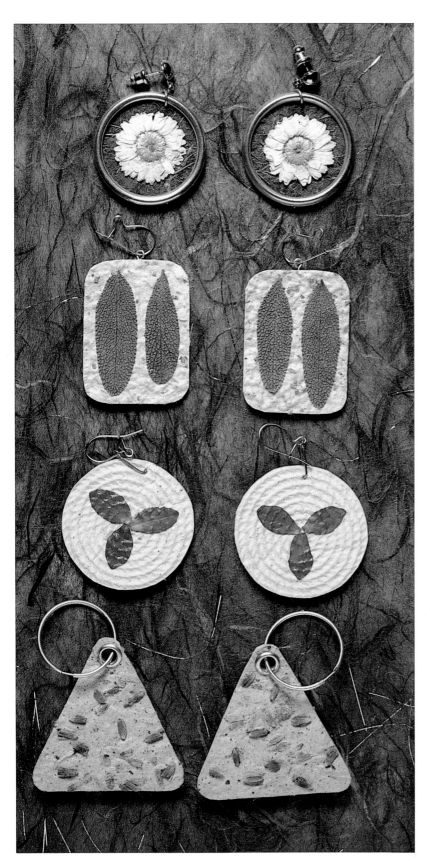

Perfumed earrings.
All the earring attachments were bought from a craft shop, the curtain rings from a hardware store. I used PVA glue to stick the bits and pieces together.

(from top to bottom)
Autumn-leaf paper, scented with camomile oil, was attached to brass curtain rings. I then glued pressed camomile flowers to the leaf paper.

Pressed sage leaves attached to sage paper.

Peppermint-scented paper impressed with coiled string. Pennyroyal mint leaves were then attached.

Lavender paper with a few extra flowers stuck on afterwards.

Using autumn leaves

Another way of producing pulp for hand-made paper is to use the actual herb or plant without additional recycled paper. Almost anything is suitable, from seaweed to straw. Here I have gathered fallen autumn leaves, mainly Norway maple, which are thin and crisp when dry, and not too thick or leathery, and then scented the final paper with essential oils of cedarwood and musk. (There does not appear to be an oil which fully captures the wonderful smell of damp woods and decomposing foliage.)

In order to break down the leaf fibre and render it suitable for paper-making, you will need to make an alkaline solution. I prefer to use natural substances rather than chemicals whenever possible, so in this case wood ash replaces the more commonly used caustic soda.

To make approximately eight sheets of 125 x 175mm (5 x 7in) paper you will need the following:

- 10-litre (2-gallon) plastic bucket
- 5-litre (1-gallon) wood ash
- 5-litre (1-gallon) water
- Rubber gloves
- Jam pan, preferably stainless steel or unchipped enamel
- A sturdy wooden stick for stirring wood ash and leaves
- A piece of finely woven net curtain approximately 70cm (28in) square
- Old plastic colander
- Food processor

Jam pan Plastic bucket Wood ash Plastic colander Rubber gloves Food processor Net curtain Wooden stick

1. Gather newly fallen leaves in a plastic bin liner. Remove the stems and any very thick veins. Put the stemless leaves into the bucket, pressing them down until it is full to the brim. Empty them on to newspaper so that the bucket can be reused.

2. In a well-ventilated kitchen and wearing rubber gloves, put the wood ash and water into the jam pan. Bring the mixture to the boil, stirring regularly to prevent the ash from sinking to the bottom of the pan and burning. When the mixture has boiled, remove it from the heat and leave it to cool, outside if possible, with the lid half on.

3. This step is best done in a sink or outside. You may also need another pair of hands to help you. Line the colander with the netting, place it over the bucket and carefully pour in the mixture. You can dispose of the strained ash in the garden.

4. Rinse the pan and netting, then re-strain the liquid back into the pan. Add the leaves and place on a high heat, stirring regularly. Bring to the boil, then cover and leave to simmer gently for ninety minutes.

5. When the liquid has cooled down, strain the mixture through the net-lined colander and rinse thoroughly with plenty of cold water.

6. Pull the net up into a ball and squeeze out the remaining water.

7. Divide the leaf mass in two, then process each separately with a little water as shown on page 12. Pour all the pulp into a bowl and add 5 litres (1 gallon) of water. Stir the mixture thoroughly.

The mixture is now ready to be made into sheets of paper. Follow the instructions on pages 14–19, remembering that the pulp has a more uneven fibre content than that of recycled paper.

8. The action of sliding in and lifting out the mould and deckle may not be quite as smooth as with recycled paper. Any pieces of fibre hanging over the edges of the deckle can be carefully removed and put back into the bowl.

9. When you are nearly at the end of the sheet-making session, when most of the pulp has been used up, hold the mould and deckle up to the light (while allowing excess water to drain away) and if there are any small areas of uncovered screen, simply scoop some pulp out of the bowl and lay it carefully over the area. The water which has contained the pulp can be strained and used as a dye (see the example on pages 60 and 62).

8

9

Taking a cutting from the autumn-leaf paper!

Autumn leaves
From left to right: autumn-leaf paper
inlaid with small pieces of dry leaf; paper
dyed with the water from the leaf pulp;
paper made with autumn-leaf pulp;
recycled computer paper with pieces of leaf
added to the pulp.

Scented calendar

Based on autumn leaves and scented with oil of cedarwood, these tinted papers were dyed with liquid from the leaf pulp. Both leaves, one pressed, the other cut from a sheet of autumn-leaf paper, were inlaid just before couching, as were the pieces of torn leaf on the tinted paper. As you can see, both papers can be written on without difficulty.

Writing paper with matching envelope

I added pieces of red autumn leaf to the pulp when making this paper (made from shredded computer paper). I also scattered some over the newly made sheets just before couching. The plain paper can be written on and would look good as an insert inside the folded inlaid paper.

O wild West Wind, thou breath of Autumn's being,
Thou, from whose unseen presence the leaves dead
Are driven, like ghosts from an enchanter fleeing,
Yellow, and black, and pale, and hectic red.

Shelley

Index